Daniel Kirkwood

Minor Planets Between Mars and Jupiter

Daniel Kirkwood

Minor Planets Between Mars and Jupiter

ISBN/EAN: 9783743372900

Manufactured in Europe, USA, Canada, Australia, Japa

Cover: Foto ©berggeist007 / pixelio.de

Manufactured and distributed by brebook publishing software
(www.brebook.com)

Daniel Kirkwood

Minor Planets Between Mars and Jupiter

THE

ASTEROIDS,

OR

MINOR PLANETS

BETWEEN

MARS AND JUPITER.

BY

DANIEL KIRKWOOD, LL.D.,

PROFESSOR EMERITUS IN THE UNIVERSITY OF INDIANA; AUTHOR OF "COMETS AND
METEORS," "METEORIC ASTRONOMY," ETC.

PHILADELPHIA:

J. B. LIPPINCOTT COMPANY.

1888.

PREFACE.

THE rapid progress of discovery in the zone of minor planets, the anomalous forms and positions of their orbits, the small size as well as the great number of these telescopic bodies, and their peculiar relations to Jupiter, the massive planet next exterior,—all entitle this part of the system to more particular consideration than it has hitherto received. The following essay is designed, therefore, to supply an obvious want. Its results are given in some detail up to the date of publication. Part I. presents in a popular form the leading historical facts as to the discovery of Ceres, Pallas, Juno, Vesta, and Astræa; a tabular statement of the dates and places of discovery for the entire group; a list of the names of discoverers, with the number of minor planets detected by each; and a table of the principal elements so far as computed.

In Part II. this descriptive summary is followed by questions relating to the origin of the cluster; the elimination of members from particular parts; the eccentricities and inclinations of the orbits; and the relation

3

of the zone to comets of short period. The elements are those given in the Paris *Annuaire* for 1887, or in recent numbers of the *Circular zum Berliner Astronomischen Jahrbuch.*

DANIEL KIRKWOOD.

Bloomington, Indiana, November, 1887.

CONTENTS.

1*

6 CONTENTS.

PART I.

THE ASTEROIDS,

OR

MINOR PLANETS BETWEEN MARS AND JUPITER.

1. Introductory.

PLANETARY DISCOVERIES BEFORE THE ASTEROIDS
WERE KNOWN.

THE first observer who watched the skies with any degree of care could not fail to notice that while the greater number of stars maintained the same relative places, a few from night to night were ever changing their positions. The planetary character of Mercury, Venus, Mars, Jupiter, and Saturn was thus known before the dawn of history. The names, however, of those who first distinguished them as " wanderers" are hopelessly lost. Venus, the morning and evening star, was long regarded as two distinct bodies. The discovery that the change of aspect was due to a single planet's change of position is ascribed to Pythagoras.

At the beginning of the seventeenth century but six primary planets and one satellite were known as members of the solar system. Very few, even of the learned, had then accepted the theory of Copernicus; in fact, before the invention of the telescope the evidence in its favor was not absolutely conclusive. On the 7th of January, 1610, Galileo first saw the satellites of Jupiter. The bearing of this discovery on the theory of the universe was sufficiently obvious. Such was the prejudice, however, against the Copernican system that some of its opponents denied even the reality of Galileo's discovery. "Those satellites," said a Tuscan astronomer, "are invisible to the naked eye, and therefore can exercise no influence on the earth, and therefore would be useless, and therefore do not exist. Besides, the Jews and other ancient nations, as well as modern Europeans, have adopted the division of the week into *seven* days, and have named them from the seven planets; now, if we increase the number of planets this whole system falls to the ground."

No other secondary planet was discovered till March 25, 1655, when Titan, the largest satellite of Saturn, was detected by Huyghens. About two years later (December 7, 1657) the same astronomer discovered the true form of Saturn's ring; and before the close of the century (1671–1684) four more satellites, Japetus, Rhea, Tethys, and Dione, were added to the Saturnian system by the elder Cassini. Our planetary system, therefore, as known at the close of the seventeenth century, consisted of six primary and ten secondary planets.

Nearly a century had elapsed from the date of Cassini's discovery of Dione, when, on the 13th of March, 1781, Sir William Herschel enlarged the dimensions of our system by the detection of a planet—Uranus— exterior to Saturn. A few years later (1787–1794) the same distinguished observer discovered the first and second satellites of Saturn, and also the four Uranian satellites. He was the only planet discoverer of the eighteenth century.

2. Discovery of the First Asteroids.

As long ago as the commencement of the seventeenth century the celebrated Kepler observed that the respective distances of the planets from the sun formed nearly a regular progression. The series, however, by which those distances were expressed required the interpolation of a term between Mars and Jupiter,—a fact which led the illustrious German to predict the discovery of a planet in that interval. This conjecture attracted but little attention till after the discovery of Uranus, whose distance was found to harmonize in a remarkable manner with Kepler's order of progression. Such a coincidence was of course regarded with considerable interest. Towards the close of the last century Professor Bode, who had given the subject much attention, published the law of distances which bears his name, but which, as he acknowledged, is due to Professor Titius. According to this formula the distances of the planets from Mercury's orbit form a geometrical series of which the ratio is two. In other words, if we reckon the distances of Venus, the earth, etc., from the orbit of Mercury, instead of from the sun, we find that—interpolating a term between Mars and Jupiter—the distance of any member of the system is very nearly half that of the next exterior. Baron De Zach, an enthusiastic astronomer, was greatly interested in Bode's empirical scheme, and undertook to determine the elements of the hypothetical planet. In 1800 a number of astronomers met at Lilienthal, organized an astronomical society, and assigned one twenty-fourth part of the zodiac to each of twenty-four observers, in order to detect, if possible, the unseen planet. When it is remembered that at this time no primary planet had been discovered within the ancient limits of the solar system, that the object to be looked for was comparatively near us, and that the so-called law of distances was purely empirical, the prospect of success, it is evident, was extremely uncertain. How long the watch, if unsuccessful, might have been continued is doubtful. The object of research, however, was fortunately brought to light before the members of the astronomical association had fairly commenced their labors.*

On the 1st of January, 1801, Professor Giuseppe Piazzi, of Palermo, noticed a star of the eighth magnitude, not indicated in Wollaston's catalogue. Subsequent observations soon revealed its planetary character, its mean distance corresponding very nearly with the calculations of De Zach. The discoverer called it Ceres Ferdinandea, in honor of his sovereign, the King of Naples. In this, however, he was not followed by astronomers, and the planet is now known by the name of Ceres alone. The discovery of this body was hailed by astronomers with the liveliest gratification as completing the harmony of the system. What, then, was their surprise when in the course of a few months this remarkable order was again interrupted! On the 28th of March, 1802, Dr. William Olbers, of Bremen, while examining the relative positions of the small stars along the path of Ceres, in order to find that planet with the greater facility, noticed a star of the seventh or eighth magnitude, forming with two others an equilateral triangle where he was certain no such configuration ex-

* The discoverer, Piazzi, was not, as has been so often affirmed, one of the astronomers to whom the search had been especially committed.

isted a few months before. In the course of a few hours its motion was perceptible, and on the following night it had very sensibly changed its position with respect to the neighboring stars. Another planet was therefore detected, and Dr. Olbers immediately communicated his discovery to Professor Bode and Baron De Zach. In his letter to the former he suggested Pallas as the name of the new member of the system,—a name which was at once adopted. Its orbit, which was soon computed by Gauss, was found to present several striking anomalies. The inclination of its plane to that of the ecliptic was nearly thirty-five degrees,—an amount of deviation altogether extraordinary. The eccentricity also was greater than in the case of any of the old planets. These peculiarities, together with the fact that the mean distances of Ceres and Pallas were nearly the same, and that their orbits approached very near each other at the intersection of their planes, suggested the hypothesis that they are fragments of a single original planet, which, at a very remote epoch, was disrupted by some mysterious convulsion. This theory will be considered when we come to discuss the tabulated elements of the minor planets now known.

For the convenience of astronomers, Professor Harding, of Lilienthal, undertook the construction of charts of all the small stars near the orbits of Ceres and Pallas. On the evening of September 1, 1804, while engaged in observations for this purpose, he noticed a star of the eighth magnitude not mentioned in the great catalogue of Lalande. This proved to be a third member of the group of asteroids. The discovery was first announced to Dr. Olbers, who observed the planet at Bremen on the evening of September 7.

Before Ceres had been generally adopted by astronomers as the name of the first asteroid, Laplace had expressed a preference for Juno. This, however, the discoverer was unwilling to accept. Mr. Harding, like Laplace, deeming it appropriate to place Juno near Jupiter, selected the name for the third minor planet, which is accordingly known by this designation.

Juno is distinguished among the first asteroids by the great eccentricity of its orbit, amounting to more than 0.25. Its least and its greatest distances from the sun are therefore to each other very nearly in the ratio of three to five. The planet consequently receives nearly three times as much light and heat in perihelion as in aphelion. It follows, also, that the half of the orbit nearest the sun is described in about eighteen months, while the remainder, or more distant half, is not passed over in much less than three years. Schroeter noticed a variation in the light of Juno, which he supposed to be produced by an axial rotation in about twenty-seven hours.

The fact that Juno was discovered not far from the point at which the orbit of Pallas approaches very near that of Ceres, was considered a strong confirmation of the hypothesis that the asteroids were produced by the explosion of a large planet; for in case this hypothesis be founded in truth, it is evident that whatever may have been the forms of the various orbits assumed by the fragments, they must all return to the point of separation. In order, therefore, to detect other members of the group, Dr. Olbers undertook a systematic examination of the two opposite regions of the heavens through which they must pass. This search was prosecuted with great industry and perseverance till ultimately crowned with success. On the 29th of March, 1807, while sweeping over one of those regions through which the orbits of the known asteroids passed, a star of the sixth magnitude was observed where none had been seen at previous examinations. Its planetary character, which was immediately suspected, was confirmed by observation, its motion being detected on the very evening of its discovery. This fortunate result afforded the first instance of the discovery of two primary planets by the same observer.

The astronomer Gauss having been requested to name the new planet, fixed upon Vesta, a name universally accepted. Though the brightest of the asteroids, its apparent diameter is too small to be accurately determined, and hence its real magnitude is not well ascertained. Professor Harrington, of Ann Arbor, has estimated the diameter at five hundred and twenty miles. According to others, however, it does not exceed three hundred. If the latter be correct, the volume is about $\frac{1}{20000}$ that of the earth. It is remarkable that notwithstanding its diminutive size it may be seen under favorable circumstances by the naked eye.

Encouraged by the discovery of Vesta (which he regarded as almost a demonstration of his favorite theory), Dr. Olbers continued his systematic search for other planetary fragments. Not meeting, however, with further success, he relinquished his observations in 1816. His failure, it may here be remarked, was doubtless owing to the fact that his examination was limited to stars of the seventh and eighth magnitudes.

The search for new planets was next resumed about 1831, by Herr Hencke, of Driessen. With a zeal and perseverance worthy of all praise, this amateur astronomer employed himself in a strict examination of the heavens represented by the Maps of the Berlin Academy. These maps extend fifteen degrees on each side of the equator, and contain all stars down to the ninth magnitude and many of the tenth. Dr. Hencke rendered some of these charts still more complete by the insertion of smaller stars; or rather, "made for himself special charts of particular districts." On the evening of December 8, 1845, he observed a star of the ninth magnitude where none had been previously seen, as he knew from the fact that it was neither found on his own chart nor given on that of the Academy. On the next morning he wrote to Professors Encke and Schumacher informing them of his supposed discovery. "It is very improbable," he remarked in his letter to the latter, "that this should prove to be merely a variable star, since in my former observations of this region, which have been continued for many years, I have never detected the slightest trace of it." The new star was soon seen at the principal observatories of Europe, and its planetary character satisfactorily established. The selection of a name was left by the discoverer to Professor Encke, who chose that of Astræa.

The facts in regard to the very numerous subsequent discoveries may best be presented in a tabular form.

TABLE I.

The Asteroids in the Order of their Discovery.

Asteroids.	Date of Discovery.	Name of Discoverer.	Place of Discovery.
1. Ceres............	1801, Jan. 1	Piazzi	Palermo
2. Pallas...........	1802, Mar. 28	Olbers	Bremen
3. Juno	1804, Sept. 1	Harding	Lilienthal
4. Vesta...........	1807, Mar. 29	Olbers	Bremen
5. Astræa...........	1845, Dec. 8	Hencke	Driessen
6. Hebe	1847, July 1	Hencke	Driessen
7. Iris	1847, Aug. 14	Hind	London
8. Flora...........	1847, Oct. 18	Hind	London
9. Metis..........	1848, Apr. 26	Graham	Markree
10. Hygeia	1849, Apr. 12	De Gasparis	Naples
11. Parthenope...	1850, May 11	De Gasparis	Naples
12. Victoria	1850, Sept. 13	Hind	London
13. Egeria..........	1850, Nov. 2	De Gasparis	Naples
14. Irene...........	1851, May 19	Hind	London
15. Eunomia......	1851, July 29	De Gasparis	Naples
16. Psyche........	1852, Mar. 17	De Gasparis	Naples
17. Thetis..........	1852, Apr. 17	Luther	Bilk
18. Melpomene...	1852, June 24	Hind	London
19. Fortuna	1852, Aug. 22	Hind	London
20. Massalia.......	1852, Sept. 19	De Gasparis	Naples
21. Lutetia........	1852, Nov. 15	Goldschmidt	Paris
22. Calliope.......	1852, Nov. 16	Hind	London
23. Thalia.........	1852, Dec. 15	Hind	London
24. Themis.........	1853, Apr. 5	De Gasparis	Naples
25. Phocea	1853, Apr. 6	Chacornac	Marseilles
26. Proserpine....	1853, May 5	Luther	Bilk
27. Euterpe........	1853, Nov. 8	Hind	London
28. Bellona	1854, Mar. 1	Luther	Bilk
29. Amphitrite ...	1854, Mar. 1	Marth	London
30. Urania.........	1854, July 22	Hind	London
31. Euphrosyne ..	1854, Sept. 1	Ferguson	Washington
32. Pomona........	1854, Oct. 26	Goldschmidt	Paris
33. Polyhymnia ..	1854, Oct. 28	Chacornac	Paris
34. Circe...........	1855, Apr. 6	Chacornac	Paris
35. Leucothea.....	1855, Apr. 19	Luther	Bilk
36. Atalanta	1855, Oct. 5	Goldschmidt	Paris
37. Fides...........	1855, Oct. 5	Luther	Bilk
38. Leda...........	1856, Jan. 12	Chacornac	Paris
39. Lætitia........	1856, Feb. 8	Chacornac	Paris
40. Harmonia.....	1856, Mar. 31	Goldschmidt	Paris
41. Daphne........	1856, May 22	Goldschmidt	Paris
42. Isis	1856, May 23	Pogson	Oxford
43. Ariadne	1857, Apr. 15	Pogson	Oxford

Table I.—*Continued.*

Asteroids.	Date of Discovery.	Name of Discoverer.	Place of Discovery.
44. Nysa............	1857, May 27	Goldschmidt	Paris
45. Eugenia	1857, June 27	Goldschmidt	Paris
46. Hestia	1857, Aug. 16	Pogson	Oxford
47. Aglaia..........	1857, Sept. 15	Luther	Bilk
48. Doris...........	1857, Sept. 19	Goldschmidt	Paris
49. Pales...........	1857, Sept. 19	Goldschmidt	Paris
50. Virginia........	1857, Oct. 4	Ferguson	Washington
51. Nemausa.......	1858, Jan. 22	Laurent	Nismes
52. Europa.........	1858, Feb. 4	Goldschmidt	Paris
53. Calypso........	1858, Apr. 4	Luther	Bilk
54. Alexandra.....	1858, Sept. 10	Goldschmidt	Paris
55. Pandora.......	1858, Sept. 10	Searle	Albany
56. Melete	1857, Sept. 9	Goldschmidt	Paris
57. Mnemosyne...	1859, Sept. 22	Luther	Bilk
58. Concordia	1860, Mar. 24	Luther	Bilk
59. Olympia.......	1860, Sept. 12	Chacornac	Paris
60. Echo............	1860, Sept. 16	Ferguson	Washington
61. Danaë	1860, Sept. 9	Goldschmidt	Paris
62. Erato...........	1860, Sept. 14	Foerster and Lesser	Berlin
63. Ausonia	1861, Feb. 10	De Gasparis	Naples
64. Angelina......	1861, Mar. 4	Tempel	Marseilles
65. Maximiliana.	1861, Mar. 8	Tempel	Marseilles
66. Maia............	1861, Apr. 9	Tuttle	Cambridge, U. S.
67. Asia............	1861, Apr. 17	Pogson	Madras
68. Leto............	1861, Apr. 29	Luther	Bilk
69. Hesperia	1861, Apr. 29	Schiaparelli	Milan
70. Panopea	1861, May 5	Goldschmidt	Paris
71. Niobe..........	1861, Aug. 13	Luther	Bilk
72. Feronia	1862, May 29	Peters and Safford	Clinton
73. Clytie.........	1862, Apr. 7	Tuttle	Cambridge
74. Galatea........	1862, Aug. 29	Tempel	Marseilles
75. Eurydice......	1862, Sept. 22	Peters	Clinton
76. Freia..........	1862, Oct. 21	D'Arrest	Copenhagen
77. Frigga.........	1862, Nov. 12	Peters	Clinton
78. Diana.	1863, Mar. 15	Luther	Bilk
79. Eurynome.....	1863, Sept. 14	Watson	Ann Arbor
80. Sappho.........	1864, May 2	Pogson	Madras
81. Terpsichore...	1864, Sept. 30	Tempel	Marseilles
82. Alcmene	1864, Nov. 27	Luther	Bilk
83. Beatrice	1865, Apr. 26	De Gasparis	Naples
84. Clio.............	1865, Aug. 25	Luther	Bilk
85. Io..............	1865, Sept. 19	Peters	Clinton
86. Semele.........	1866, Jan. 14	Tietjen	Berlin
87. Sylvia	1866, May 16	Pogson	Madras
88. Thisbe.........	1866, June 15	Peters	Clinton
89. Julia...........	1866, Aug. 6	Stephan	Marseilles
90. Antiope........	1866, Oct. 1	Luther	Bilk
91. Ægina.........	1866, Nov. 4	Borelly	Marseilles
92. Undina........	1867, July 7	Peters	Clinton

Table I.—*Continued.*

Asteroids.	Date of Discovery.	Name of Discoverer.	Place of Discovery.
93. Minerva.......	1867, Aug. 24	Watson	Ann Arbor
94. Aurora.........	1867, Sept. 6	Watson	Ann Arbor
95. Arethusa......	1867, Nov. 24	Luther	Bilk
96. Ægle....	1868, Feb. 17	Coggia	Marseilles
97. Clotho	1868, Feb. 17	Coggia	Marseilles
98. Ianthe..........	1868, Apr. 18	Peters	Clinton
99. Dike	1868, May 28	Borelly	Marseilles
100. Hecate........	1868, July 11	Watson	Ann Arbor
101. Helena	1868, Aug. 15	Watson	Ann Arbor
102. Miriam........	1868, Aug. 22	Peters	Clinton
103. Hera...	1868, Sept. 7	Watson	Ann Arbor
104. Clymene	1868, Sept. 13	Watson	Ann Arbor
105. Artemis.......	1868, Sept. 16	Watson	Ann Arbor
106. Dione	1868, Oct. 10	Watson	Ann Arbor
107. Camilla.......	1868, Nov. 17	Pogson	Madras
108. Hecuba........	1869, Apr. 2	Luther	Bilk
109. Felicitas	1869, Oct. 9	Peters	Clinton
110. Lydia..........	1870, Apr. 19	Borelly	Marseilles
111. Ate	1870, Aug. 14	Peters	Clinton
112. Iphigenia. ...	1870, Sept. 19	Peters	Clinton
113. Amalthea......	1871, Mar. 12	Luther	Bilk
114. Cassandra.....	1871, July 23	Peters	Clinton
115. Thyra..........	1871, Aug. 6	Watson	Ann Arbor
116. Sirona	1871, Sept. 8	Peters	Clinton
117. Lomia	1871, Sept. 12	Borelly	Marseilles
118. Peitho	1872, Mar. 15	Luther	Bilk
119. Althea..........	1872, Apr. 3	Watson	Ann Arbor
120. Lachesis.......	1872, Apr. 10	Borelly	Marseilles
121. Hermione.....	1872, May 12	Watson	Ann Arbor
122. Gerda	1872, July 31	Peters	Clinton
123. Brunhilda	1872, July 31	Peters	Clinton
124. Alceste........	1872, Aug. 23	Peters	Clinton
125. Liberatrix....	1872, Sept. 11	Prosper Henry	Paris
126. Velleda	1872, Nov. 5	Paul Henry	Paris
127. Johanna	1872, Nov. 5	Prosper Henry	Paris
128. Nemesis.......	1872, Nov. 25	Watson	Ann Arbor
129. Antigone......	1873, Feb. 5	Peters	Clinton
130. Electra........	1873, Feb. 17	Peters	Clinton
131. Vala...........	1873, May 24	Peters	Clinton
132. Æthra	1873, June 13	Watson	Ann Arbor
133. Cyrene........	1873, Aug. 16	Watson	Ann Arbor
134. Sophrosyne...	1873, Sept. 27	Luther	Bilk
135. Hertha........	1874, Feb. 18	Peters	Clinton
136. Austria	1874, Mar. 18	Palisa	Pola
137. Meliboea	1874, Apr. 21	Palisa	Pola
138. Tolosa	1874, May 19	Perrotin	Toulouse
139. Juewa	1874, Oct. 10	Watson	Pekin
140. Siwa...........	1874, Oct. 13	Palisa	Pola
141. Lumen........	1875, Jan. 13	Paul Henry	Paris

Table I.—*Continued.*

Asteroids.	Date of Discovery.	Name of Discoverer.	Place of Discovery.
142. Polana.........	1875, Jan. 28	Palisa	Pola
143. Adria.	1875, Feb. 23	Palisa	Pola
144. Vibilia..........	1875, June 3	Peters	Clinton
145. Adeona	1875, June 3	Peters	Clinton
146. Lucina..........	1875, June 8	Borelly	Marseilles
147. Protogenea ...	1875, July 10	Schulhof	Vienna
148. Gallia	1875, Aug. 7	Prosper Henry	Paris
149. Medusa..	1875, Sept. 21	Perrotin	Toulouse
150. Nuwa	1875, Oct. 18	Watson	Ann Arbor
151. Abundantia...	1875, Nov. 1	Palisa	Pola
152. Atala............	1875, Nov. 2	Paul Henry	Paris
153. Hilda..........	1875, Nov. 2	Palisa	Pola
154. Bertha.........	1875, Nov. 4	Prosper Henry	Paris
155. Scylla..........	1875, Nov. 8	Palisa	Pola
156. Xantippe......	1875, Nov. 22	Palisa	Pola
157. Dejanira	1875, Dec. 1	Borelly	Marseilles
158. Coronis........	1876, Jan. 4	Knorre	Berlin
159. Æmilia.........	1876, Jan. 26	Paul Henry	Paris
160. Una	1876, Feb. 20	Peters	Clinton
161. Athor..........	1876, Apr. 19	Watson	Ann Arbor
162. Laurentia.....	1876, Apr. 21	Prosper Henry	Paris
163. Erigone........	1876, Apr. 26	Perrotin	Toulouse
164. Eva.............	1876, July 12	Paul Henry	Paris
165. Loreley	1876, Aug. 9	Peters	Clinton
166. Rhodope.......	1876, Aug. 15	Peters	Clinton
167. Urda...........	1876, Aug. 28	Peters	Clinton
168. Sibylla.........	1876, Sept. 27	Watson	Ann Arbor
169. Zelia............	1876, Sept. 28	Prosper Henry	Paris
170. Maria...	1877, Jan. 10	Perrotin	Toulouse
171. Ophelia........	1877, Jan. 13	Borelly	Marseilles
172. Baucis..........	1877, Feb. 5	Borelly	Marseilles
173. Ino..............	1877, Aug. 1	Borelly	Marseilles
174. Phædra.........	1877, Sept. 2	Watson	Ann Arbor
175. Andromache..	1877, Oct. 1	Watson	Ann Arbor
176. Idunna.........	1877, Oct. 14	Peters	Clinton
177. Irma...........	1877, Nov. 5	Paul Henry	Paris
178. Belisana.......	1877, Nov. 6	Palisa	Pola
179. Clytemnestra.	1877, Nov. 11	Watson	Ann Arbor
180. Garumna.......	1878, Jan. 29	Perrotin	Toulouse
181. Eucharis......	1878, Feb. 2	Cottenot	Marseilles
182. Elsa............	1878, Feb. 7	Palisa	Pola
183. Istria...........	1878, Feb. 8	Palisa	Pola
184. Deiopea........	1878, Feb. 28	Palisa	Pola
185. Eunice.........	1878, Mar. 1	Peters	Clinton
186. Celuta.........	1878, Apr. 6	Prosper Henry	Paris
187. Lamberta......	1878, Apr. 11	Coggia	Marseilles
188. Menippe.......	1878, June 18	Peters	Clinton
189. Phthia.........	1878, Sept. 9	Peters	Clinton
190. Ismene.........	1878, Sept. 22	Peters	Clinton

Table I.—*Continued.*

Asteroids.	Date of Discovery.	Name of Discoverer.	Place of Discovery.
191. Kolga............	1878, Sept. 30	Peters	Clinton
192. Nausicaa	1879, Feb. 17	Palisa	Pola
193. Ambrosia......	1879, Feb. 28	Coggia	Marseilles
194. Procne.........	1879, Mar. 21	Peters	Clinton
195. Euryclea.......	1879, Apr. 22	Palisa	Pola
196. Philomela.....	1879, May 14	Peters	Clinton
197. Arete...........	1879, May 21	Palisa	Pola
198. Ampella........	1879, June 13	Borelly	Marseilles
199. Byblis..........	1879, July 9	Peters	Clinton
200. Dynamene.....	1879, July 27	Peters	Clinton
201. Penelope.......	1879, Aug. 7	Palisa	Pola
202. Chryseis........	1879, Sept. 11	Peters	Clinton
203. Pompeia.......	1879, Sept. 25	Peters	Clinton
204. Callisto........	1879, Oct. 8	Palisa	Pola
205. Martha.........	1879, Oct. 13	Palisa	Pola
206. Hersilia........	1879, Oct. 13	Peters	Clinton
207. Hedda..........	1879, Oct. 17	Palisa	Pola
208. Isabella........	1879, Oct. 21	Palisa	Pola
209. Dido	1879, Oct. 22	Peters	Clinton
210. Lachrymosa ..	1879, Nov. 12	Palisa	Pola
211. Isolda..........	1879, Dec. 10	Palisa	Pola
212. Medea	1880, Feb. 6	Palisa	Pola
213. Lilæa..........	1880, Feb. 16	Peters	Clinton
214. Aschera.......	1880, Feb. 26	Palisa	Pola
215. Œnone	1880, Apr. 7	Knorre	Berlin
216. Cleopatra......	1880, Apr. 10	Palisa	Pola
217. Eudora.........	1880, Aug. 30	Coggia	Marseilles
218. Bianca.........	1880, Sept. 4	Palisa	Pola
219. Thusnelda.....	1880, Sept. 20	Palisa	Pola
220. Stephania......	1881, May 19	Palisa	Vienna
221. Eos.............	1882, Jan. 18	Palisa	Vienna
222. Lucia..........	1882, Feb. 9	Palisa	Vienna
223. Rosa...........	1882, Mar. 9	Palisa	Vienna
224. Oceana.........	1882, Mar. 30	Palisa	Vienna
225. Henrietta......	1882, Apr. 19	Palisa	Vienna
226. Weringia	1882, July 19	Palisa	Vienna
227. Philosophia...	1882, Aug. 12	Paul Henry	Paris
228. Agathe.........	1882, Aug. 19	Palisa	Vienna
229. Adelinda.......	1882, Aug. 22	Palisa	Vienna
230. Athamantis...	1882, Sept. 3	De Ball	Bothcamp
231. Vindobona....	1882, Sept. 10	Palisa	Vienna
232. Russia.........	1883, Jan. 31	Palisa	Vienna
233. Asterope.......	1883, May 11	Borelly	Marseilles
234. Barbara........	1883, Aug. 13	Peters	Clinton
235. Caroline........	1883, Nov. 29	Palisa	Vienna
236. Honoria........	1884, Apr. 26	Palisa	Vienna
237. Cœlestina......	1884, June 27	Palisa	Vienna
238. Hypatia........	1884, July 1	Knorre	Berlin
239. Adrastea.......	1884, Aug. 18	Palisa	Vienna

Table I.—*Continued.*

Asteroids.	Date of Discovery.	Name of Discoverer.	Place of Discovery.
240. Vanadis.........	1884, Aug. 27	Borelly	Marseilles
241. Germania......	1884, Sept. 12	Luther	Dusseldorf
242. Krïemhild.....	1884, Sept. 22	Palisa	Vienna
243. Ida...............	1884, Sept. 29	Palisa	Vienna
244. Sita..............	1884, Oct. 14	Palisa	Vienna
245. Vera..............	1885, Feb. 6	Pogson	Madras
246. Asporina.......	1885, Mar. 6	Borelly	Marseilles
247. Eukrate.........	1885, Mar. 14	Luther	Dusseldorf
248. Lameia.........	1885, June 5	Palisa	Vienna
249. Ilse..............	1885, Aug. 17	Peters	Clinton
250. Bettina.........	1885, Sept. 3	Palisa	Vienna
251. Sophia...........	1885, Oct. 4	Palisa	Vienna
252. Clementina....	1885, Oct. 27.	Perrotin	Nice
253. Mathilda.......	1885, Nov. 12	Palisa	Vienna
254. Augusta........	1886, Mar. 31	Palisa	Vienna
255. Oppavia........	1886, Mar. 31	Palisa	Vienna
256. Walpurga......	1886, Apr. 3	Palisa	Vienna
257. Silesia...........	1886, Apr. 5	Palisa	Vienna
258. Tyche...........	1886, May 4	Luther	Dusseldorf
259. Altheia.........	1886, June 28	Peters	Clinton
260. Huberta.......	1886, Oct. 3	Palisa	Vienna
261. Prymno.........	1886, Oct. 31	Peters	Clinton
262. Valda...........	1886, Nov. 3	Palisa	Vienna
263. Dresda..........	1886, Nov. 3	Palisa	Vienna
264. Libussa.........	1886, Dec. 17	Peters	Clinton
265. Anna............	1887, Feb. 25	Palisa	Vienna
266. Aline............	1887, May 17	Palisa	Vienna
267. Tirza............	1887, May 27	Charlois	Nice
268.	1887, June 9	Borelly	Marseilles.
269.	1887, Sept. 21	Palisa	Vienna
270.	1887, Oct. 8	Peters	Clinton
271.	1887, Oct. 16	Knorre	Berlin

3. Remarks on Table I.

The numbers discovered by the thirty-five observers are respectively as follows:

Palisa	60
Peters	47
Luther	23
Watson	22
Borelly	15
Goldschmidt	14
Hind	10
De Gasparis	9
Pogson	8
Paul Henry	7
Prosper Henry	7
Chacornac	6
Perrotin	6
Coggia	5
Knorre	4
Tempel	4
Ferguson	3
Olbers	2
Hencke	2
Tuttle	2
Foerster (with Lesser)	1
Safford (with Peters)	1
and Messrs. Charlois, Cottenot, D'Arrest, De Ball, Graham, Harding, Laurent, Piazzi, Schiaparelli, Schulhof, Stephan, Searle, and Tietjen, each	1

Before arrangements had been made for the telegraphic transmission of discoveries between Europe and America, or even between the observatories of Europe, the same planet was sometimes independently discovered by different observers. For example, Virginia was found by Ferguson, at Washington, on October 4, 1857, and by Luther, at Bilk, fifteen days later.   In all cases, however, credit has been given to the first observer.

Hersilia, the two hundred and sixth of the group, was lost before sufficient observations were obtained for determining its elements.   It was not rediscovered till December 14, 1884.   Menippe, the one hundred and eighty-eighth, was also lost soon after its discovery in 1878.   It has not been seen for more than nine years, and considerable uncertainty attaches to its estimated elements.

Of the two hundred and seventy-one members now known (1887), one hundred and ninety-one have been discovered in Europe, seventy-four in America, and six in Asia.   The years of most successful search, together with the number discovered in each, were:

	Asteroids.
1879	20
1875	17
1868	12
1878	12

And six has been the average yearly number since the commencement of renewed effort in 1845.   All the larger members of the group have, doubtless, been discovered.   It seems not improbable, however, that an indefinite number of very small bodies belonging to the zone remain to be found.   The process of discovery is becoming more difficult as the known number increases. The astronomer, for instance, who may discover number two hundred and seventy-two must know the simultaneous positions of the two hundred and seventy-one previously detected before he can decide whether he has picked up a new planet or merely rediscovered an old one.   The numbers discovered in the several months are as follows:

January	13	July	14
February	23	August	28
March	19	September	46
April	35	October	28
May	21	November	26
June	13	December	5

This obvious disparity is readily explained. The weather is favorable for night watching in April and September; the winter months are too cold for continuous observations; and the small numbers in June and July may be referred to the shortness of the nights.

4. Mode of Discovery.

The astronomer who would undertake the search for new asteroids must supply himself with star-charts extending some considerable distance on each side of the ecliptic, and containing all telescopic stars down to the thirteenth or fourteenth magnitude. The detection of a star not found in the chart of a particular section will indicate its motion, and hence its planetary character. The construction of such charts has been a principal object in the labors of Dr. Peters, at Clinton, New York. In fact, his discovery of minor planets has in most instances been merely an incidental result of his larger and more important work.

NAMES AND SYMBOLS.

The fact that the names of female deities in the Greek and Roman mythologies had been given to the first asteroids suggested a similar course in the selection of names after the new epoch of discovery in 1845. While conformity to this rule has been the general aim

3

of discoverers, the departures from it have been increasingly numerous. The twelfth asteroid, discovered in London, was named Victoria, in honor of the reigning sovereign; the twentieth and twenty-fifth, detected at Marseilles,* received names indicative of the place of their discovery; Lutetia, the first found at Paris, received its name for a similar purpose; the fifty-fourth was named Alexandra, for Alexander von Humboldt; the sixty-seventh, found by Pogson at Madras, was named Asia, to commemorate the fact that it was the first discovered on that continent. We find, also, Julia, Bertha, Xantippe, Zelia, Maria, Isabella, Martha, Dido, Cleopatra, Barbara, Ida, Augusta, and Anna. Why these were selected we will not stop to inquire.

As the number of asteroids increased it was found inconvenient to designate them individually by particular signs, as in the case of the old planets. In 1849, Dr. B. A. Gould proposed to represent them by the numbers expressing their order of discovery enclosed in a small circle. This method was at once very generally adopted.

5. Magnitudes of the Asteroids.

The apparent diameter of the largest is less than one-second of arc. They are all too small, therefore, to be accurately measured by astronomical instruments. From photometric observations, however, Argelander,† Stone,‡ and Pickering § have formed estimates of the diameters,

---

* Massalia was discovered by De Gasparis, at Naples, Sept. 19, 1852, and independently, the next night, by Chacornac, at Marseilles. The name was given by the latter.

† Astr. Nach., No. 932.

‡ Monthly Notices, vol. xxvii.

§ Annals of the Obs. of Harv. Coll., 1879.

the results giving probably close approximations to the true magnitudes. According to these estimates the diameter of the largest, Vesta, is about three hundred miles, that of Ceres about two hundred, and those of Pallas and Juno between one and two hundred. The diameters of about thirty are between fifty and one hundred miles, and those of all others less than fifty; the estimates for Menippe and Eva giving twelve and thirteen miles respectively. The diameter of the former is to that of the earth as one to six hundred and sixty-four; and since spheres are to each other as the cubes of their diameters, it would require two hundred and ninety millions of such asteroids to form a planet as large as our globe. In other words, if the earth be represented by a sphere one foot in diameter, the magnitude of Menippe on the same scale would be that of a sand particle whose diameter is one fifty-fifth of an inch. Its surface contains about four hundred and forty square miles,—an area equal to a county twenty-one miles square. The surface attractions of two planets having the same density are to each other as their diameters. A body, therefore, weighing two hundred pounds at the earth's surface would on the surface of the asteroid weigh less than five ounces. At the earth's surface a weight falls sixteen feet the first second, at the surface of Menippe it would fall about one-fourth of an inch. A person might leap from its surface to a height of several hundred feet, in which case he could not return in much less than an hour. "But of such speculations," Sir John Herschel remarks, "there is no end."

The number of these planetules between the orbits of Mars and Jupiter in all probability can never be known. It was estimated by Leverrier that the quantity of matter contained in the group could not be greater than one-fourth of the earth's mass.   But this would be equal to five thousand planets, each as large as Vesta, to seventy-two millions as large as Menippe, or to four thousand millions of five miles in diameter.   In short, the existence of an indefinite number too small for detection by the most powerful glasses is by no means improbable. The more we study this wonderful section of the solar system, the more mystery seems to envelop its origin and constitution.

6. The Orbits of the Asteroids.

The form, magnitude, and position of a planet's orbit are determined by the following elements :

1. The semi-axis major, or mean distance, denoted by the symbol a.

2. The eccentricity, e.

3. The longitude of the perihelion, π.

4. The longitude of the ascending node, Ω.

5. The inclination, or the angle contained between the plane of the orbit and that of the ecliptic, i.

And in order to compute a planet's place in its orbit for any given time we must also know

6. Its period, P, and

7. Its mean longitude, l, at a given epoch.

These elements, except the last, are given for all the asteroids, so far as known, in Table II.   In column first the number denoting the order of discovery is attached to each name.

TABLE II.

Elements of the Asteroids.

Name	a	P	e	π		Ω		i	
149. Medusa...........	2.1327	1137.7 d	0.1194	246°	37'	342°	13'	1°	6'
244. Sita...............	2.1765	1172.8	0.1370	13	8	208	37	2	50
228. Agathe...........	2.2009	1192.6	0.2405	329	23	313	18	2	33
8. Flora.............	2.2014	1193.3	0.1567	32	54	110	18	5	53
43. Ariadne.........	2.2033	1194.5	0.1671	277	58	264	35	3	28
254. Augusta.........	2.2060	1196.8	0.1227	260	47	28	9	4	36
72. Feronia.........	2.2661	1246.0	0.1198	307	58	207	49	5	24
40. Harmonia.......	2.2673	1247.0	0.0466	0	54	93	35	4	16
207. Hedda............	2.2839	1260.7	0.0301	217	2	28	51	3	49
136. Austria...........	2.2863	1262.7	0.0849	316	6	186	7	9	33
18. Melpomene.....	2.2956	1270.4	0.2177	15	6	150	4	10	9
80. Sappho..........	2.2962	1270.9	0.2001	355	18	218	44	8	37
261. Prymno..........	2.3062	1278.4	0.0794	179	35	96	33	3	38
12. Victoria..........	2.3342	1302.7	0.2189	301	39	235	35	8	23
27. Euterpe..........	2.3472	1313.5	0.1739	87	59	93	51	1	36
219. Thusnelda......	2.3542	1319.4	0.2247	340	34	200	44	10	47
163. Erigone..........	2.3560	1320.9	0.1567	93	46	159	2	4	42
169. Zelia	2.3577	1322.3	0.1313	326	20	354	38	5	31
4. Vesta	2.3616	1325.6	0.0884	250	57	103	29	7	8
186. Celuta............	2.3623	1326.2	0.1512	327	24	14	34	13	6
84. Clio................	2.3629	1326.7	0.2360	339	20	327	28	9	22
51. Nemausa	2.3652	1328.6	0.0672	174	43	175	52	9	57
220. Stephania.......	2.3666	1329.8	0.2653	332	53	258	24	7	35
30. Urania..........	2.3667	1329.9	0.1266	31	46	308	12	2	6
105. Artemis	2.3744	1336.4	0.1749	242	38	188	3	21	31
113. Amalthea.......	2.3761	1337.8	0.0874	198	44	123	11	5	2
115. Thyra............	2.3791	1340.3	0.1939	43	2	309	5	11	35
161. Athor............	2.3792	1340.5	0.1389	310	40	18	27	9	3
172. Baucis	2.3794	1340.6	0.1139	329	23	331	50	10	2
249. Ilse......	2.3795	1340.6	0.2195	14	17	334	49	9	40
230. Athamantis	2.3842	1344.6	0.0615	17	31	239	33	9	26
7. Iris...............	2.3862	1346.4	0.2308	41	23	259	48	5	28
9. Metis.......	2.3866	1346.7	0.1233	71	4	68	32	5	36
234. Barbara.........	2.3873	1347.3	0.2440	333	26	144	9	15	22
60. Echo	2.3934	1352.4	0.1838	98	36	192	5	3	35
63. Ausonia.........	2.3979	1356.3	0.1239	270	25	337	58	5	48
25. Phocea	2.4005	1358.5	0.2553	302	48	208	27	21	35
192. Nausicaa........	2.4014	1359.3	0.2413	343	19	160	46	6	50
20. Massalia........	2.4024	1365.8	0.1429	99	7	206	36	0	41
265. Anna............	2.4096	1366.2	0.2628	226	18	335	26	25	24
182. Elsa	2.4157	1371.4	0.1852	51	52	106	30	2	0
142. Polana..........	2.4194	1374.5	0.1322	219	54	317	34	2	14
67. Asia..............	2.4204	1375.4	0.1866	306	35	202	47	5	59
44. Nysa	2.4223	1377.0	0.1507	111	57	131	11	3	42

3*

Table II.—*Continued*.

Name	*a*	*P*	*e*	*π*		*Ω*		*i*	
6. Hebe	2.4254	1379.3 d	0.2034	15°	16'	138°	43'	10°	47'
83. Beatrix	2.4301	1383.6	0.0859	191	46	27	32	5	0
135. Hertha	2.4303	1383.8	0.2037	320	11	344	3	2	19
131. Vala	2.4318	1385.1	0.0683	222	50	65	15	4	58
112. Iphigenia	2.4335	1386.6	0.1282	338	9	324	3	2	37
21. Lutetia	2.4354	1388.2	0.1621	327	4	80	28	3	5
118. Peitho	2.4384	1390.8	0.1608	77	36	47	30	7	48
126. Velledo	2.4399	1392.1	0.1061	347	46	23	7	2	56
42. Isis	2.4401	1392.2	0.2256	317	58	84	28	8	35
19. Fortuna	2.4415	1394.4	0.1594	31	3	211	27	1	33
79. Eurynome	2.4436	1395.2	0.1945	44	22	206	44	4	37
138. Tolosa	2.4492	1400.0	0.1623	311	39	54	52	3	14
189. Phthia	2.4505	1401.1	0.0356	6	50	203	22	5	10
11. Parthenope	2.4529	1403.2	0.0994	318	2	125	11	4	37
178. Belisana	2.4583	1407.8	0.1266	278	0	50	17	2	5
198. Ampella	2.4595	1408.9	0.2266	354	46	268	45	9	20
248. Lameia	2.4714	1419.1	0.0656	248	40	246	34	4	1
17. Thetis	2.4726	1420.1	0.1293	261	37	125	24	5	36
46. Hestia	2.5265	1466.8	0.1642	354	14	181	31	2	17
89. Julia	2.5510	1488.2	0.1805	353	13	311	42	16	11
232. Russia	2.5522	1489.3	0.1754	200	25	152	30	6	4
29. Amphitrite	2.5545	1491.3	0.0742	56	23	356	41	6	7
170. Maria	2.5549	1491.7	0.0639	95	47	301	20	14	23
262. Valda	2.5635	1496.4	0.2172	61	42	38	40	7	46
258. Tyche	2.5643	1499.8	0.1966	15	42	208	4	14	50
134. Sophrosyne	2.5647	1500.3	0.1165	67	33	346	22	11	36
264. Libussa	2.5672	1502.4	0.0925	0	7	50	23	10	29
193. Ambrosia	2.5758	1510.0	0.2854	70	52	351	15	11	39
13. Egeria	2.5765	1510.6	0.0871	120	10	43	12	16	32
5. Astræa	2.5786	1512.4	0.1863	134	57	141	28	5	19
119. Althea	2.5824	1515.7	0.0815	11	29	203	57	5	45
157. Dejanira	2.5828	1516.1	0.2105	107	24	62	31	12	2
101. Helena	2.5849	1518.0	0.1386	327	15	343	46	10	11
32. Pomona	2.5873	1520.1	0.0830	193	22	220	43	5	29
91. Ægina	2.5895	1522.1	0.1087	80	22	11	7	2	8
14. Irene	2.5896	1522.1	0.1627	180	19	86	48	9	8
111. Ate	2.5927	1524.8	0.1053	108	42	306	13	4	57
151. Abundantia	2.5932	1525.3	0.0356	173	55	38	48	6	30
56. Melete	2.6010	1532.2	0.2340	294	50	194	1	8	2
132. Æthra	2.6025	1533.5	0.3799	152	24	260	2	25	0
214. Aschera	2.6111	1541.1	0.0316	115	55	342	30	3	27
70. Panopea	2.6139	1543.6	0.1826	299	49	48	18	11	38
194. Procne	2.6159	1545.4	0.2383	319	33	159	19	18	24
53. Calypso	2.6175	1546.8	0.2060	92	52	143	58	5	7
78. Diana	2.6194	1548.5	0.2088	121	42	333	58	8	40
124. Alceste	2.6297	1557.6	0.0784	245	42	188	26	2	56
23. Thalia	2.6306	1558.4	0.2299	123	58	67	45	10	14
164. Eva	2.6314	1559.1	0.3471	359	32	77	28	24	25
15. Eunomia	2.6437	1570.0	0.1872	27	52	188	26	2	56
37. Fides	2.6440	1570.3	0.1758	66	26	8	21	3	7

Table II.—*Continued.*

Name	a	P	e	π		Ω		i	
66. Maia	2.6454	1571.6 d	0.1750	48°	8′	8°	17′	3°	6′
224. Oceana	2.6465	1572.6	0.0455	270	51	353	18	5	52
253. Mathilde	2.6469	1572.9	0.2620	333	39	180	3	6	37
50. Virginia	2.6520	1577.4	0.2852	10	9	173	45	2	48
144. Vibilia	2.6530	1578.4	0.2348	7	9	76	47	4	48
85. Io	2.6539	1579.2	0.1911	322	35	203	56	11	53
26. Proserpine	2.6561	1581.1	0.0873	236	25	45	55	3	36
233. Asterope	2.6596	1584.3	0.1010	344	36	222	25	7	39
102. Miriam	2.6619	1586.3	0.3035	354	39	211	58	5	4
240. Venadis	2.6638	1588.0	0.2056	51	53	114	54	2	6
73. Clytie	2.6652	1589.3	0.0419	57	55	7	51	2	24
218. Bianca	2.6653	1589.3	0.1155	230	14	170	50	15	13
141. Lumen	2.6666	1590.5	0.2115	13	43	319	7	11	57
77. Frigga	2.6680	1591.8	0.1318	58	47	2	0	2	28
3. Juno	2.6683	1592.0	0.2579	54	50	170	53	13	1
97. Clotho	2.6708	1594.3	0.2550	65	32	160	37	11	46
75. Eurydice	2.6720	1595.3	0.3060	335	33	359	56	5	1
145. Adeona	2.6724	1595.4	0.1406	117	53	77	41	12	38
204. Callisto	2.6732	1596.4	0.1752	257	45	205	40	8	19
114. Cassandra	2.6758	1598.8	0.1401	153	6	164	24	4	55
201. Penelope	2.6764	1599.3	0.1818	334	21	157	5	5	44
64. Angelina	2.6816	1603.9	0.1271	125	36	311	4	1	19
98. Ianthe	2.6847	1606.7	0.1920	148	52	354	7	15	32
34. Circe	2.6864	1608.3	0.1073	148	41	184	46	5	27
123. Brunhilda	2.6918	1613.2	0.1150	72	57	308	28	6	27
166. Rhodope	2.6927	1613.9	0.2140	30	51	129	33	12	2
109. Felicitas	2.6950	1616.0	0.3002	56	1	4	56	8	3
246. Asporina	2.6994	1619.9	0.1065	255	54	162	35	15	39
58. Concordia	2.7004	1620.8	0.0426	189	10	161	20	5	2
103. Hera	2.7014	1621.8	0.0803	321	3	136	18	5	24
54. Alexandra	2.7095	1629.1	0.2000	295	39	313	45	11	47
226. Weringia	2.7118	1631.2	0.2048	284	46	135	18	15	50
59. Olympia	2.7124	1631.7	0.1189	17	33	170	26	8	37
146. Lucina	2.7189	1637.5	0.0655	227	34	84	16	13	6
45. Eugenia	2.7205	1639.0	0.0811	232	5	147	57	6	35
210. Isabella	2.7235	1641.7	0.1220	44	22	32	58	5	18
187. Lamberta	2.7272	1645.0	0.2391	214	4	22	13	10	43
180. Garumna	2.7286	1646.3	0.1722	125	56	314	42	0	54
160. Una	2.7287	1646.4	0.0624	55	57	9	22	3	51
140. Siwa	2.7316	1649.0	0.2160	300	33	107	2	3	12
110. Lydia	2.7327	1650.0	0.0770	336	49	57	10	6	0
185. Eunice	2.7372	1654.1	0.1292	16	32	153	50	23	17
203. Pompeia	2.7376	1654.5	0.0588	42	51	348	37	3	13
200. Dynamene	2.7378	1654.6	0.1335	46	38	325	26	6	56
197. Arete	2.7390	1655.8	0.1621	324	51	82	6	8	48
206. Hersilia	2.7399	1656.5	0.0389	95	44	145	16	3	46
255. Oppavia	2.7402	1656.6	0.0728	169	15	14	6	9	33
247. Eukrate	2.7412	1657.7	0.2387	53	44	0	20	25	7
38. Leda	2.7432	1659.6	0.1531	101	20	296	27	6	57
125. Liberatrix	2.7437	1660.0	0.0798	273	29	169	35	4	38

Table II.—*Continued.*

Name	a	P	e	π		Ω		i	
173. Ino.................	2.7446	1660.8 d	0.2047	13°	28′	148°	34′	14°	15′
36. Atalanta.........	2.7452	1661.3	0.3023	42	44	359	14	18	42
128. Nemesis.........	2.7514	1666.9	0.1257	16	34	76	31	6	16
93. Minerva.........	2.7537	1669.0	0.1405	274	44	5	4	8	37
177. Johanna.........	2.7550	1670.3	0.0659	122	37	31	46	8	17
71. Niobe............	2.7558	1671.0	0.1732	221	17	316	30	23	19
?3. Lilæa............	2.7563	1671.4	0.1437	281	4	122	17	6	47
65. Pandora.........	2.7604	1675.1	0.1429	10	36	10	56	7	14
237. Celestina..	2.7607	1675.5	0.0738	282	49	84	33	9	46
143. Adria............	2.7619	1676.6	0.0729	222	27	333	42	11	30
82. Alcmene.........	2.7620	1676.6	0.2228	131	45	26	57	2	51
116. Sirona	2.7669	1681.1	0.1433	152	47	64	26	3	35
1. Ceres............	2.7673	1681.4	0.0763	149	38	80	47	10	37
88. Thisbe...........	2.7673	1681.5	0.1632	308	34	277	54	16	11
215. Œnone	2.7679	1682.0	0.0390	346	24	25	25	1	44
2. Pallas...........	2.7680	1682.1	0.2408	122	12	172	45	34	44
39. Lætitia.........	2.7680	1682.1	0.1142	3	8	157	15	10	22
41. Daphne	2.7688	1682.8	0.2674	220	33	179	8	15	58
177. Irma	2.7695	1683.5	0.2370	22	6	349	17	1	27
148. Gallia...........	2.7710	1684.8	0.1855	36	7	145	13	25	21
267. Tirza............	2.7742	1687.6	0.0986	264	5	73	59	6	2
74. Galatea.........	2.7770	1690.3	0.2392	8	18	197	51	4	0
205. Martha.	2.7771	1690.4	0.1752	21	54	212	12	10	40
139. Juewa...........	2.7793	1692.4	0.1773	164	34	2	21	10	57
28. Bellona..........	2.7797	1692.7	0.1491	124	1	144	37	9	22
68. Leto	2.7805	1693.5	0.1883	345	14	45	1	7	58
216. Cleopatra.......	2.7964	1708.0	0.2492	328	15	215	49	13	2
99. Dike	2.7966	1708.3	0.2384	240	36	41	44	13	53
236. Honoria.........	2.7993	1710.7	0.1893	356	59	186	27	7	37
183. Istria	2.8024	1713.4	0.3530	45	0	142	46	26	33
266. Aline............	2.8078	1718.5	0.1573	23	52	236	18	13	20
188. Menippe.........	2.8211	1730.7	0.2173	309	38	241	44	11	21
167. Urda.	2.8533	1760.4	0.0340	296	4	166	28	2	11
81. Terpsichore.....	2.8580	1764.8	0.2080	49	1	2	25	7	55
174. Phedra	2.8600	1766.6	0.1492	253	12	328	49	12	9
243. Ida...............	2.8610	1767.5	0.0419	71	22	326	21	1	10
242. Kriemhild......	2.8623	1768.7	0.1219	123	1	207	57	11	17
129. Antigone........	2.8678	1773.9	0.2126	242	4	137	37	12	10
217. Eudora	2.8690	1774.9	0.3068	314	41	164	10	10	19
158. Coronis	2.8714	1777.2	0.0545	56	56	281	30	1	0
33. Polyhymnia....	2.8751	1780.7	0.3349	342	59	9	19	1	56
195. Euryclea........	2.8790	1784.2	0.0471	115	48	7	57	7	1
235. Caroline	2.8795	1784.7	0.0595	268	29	66	35	9	4
47. Aglaia	2.8819	1786.9	0.1317	312	40	40	20	5	1
208. Lachrymosa....	2.8926	1796.9	0.0149	127	52	5	43	1	48
191. Kolga...........	2.8967	1800.8	0.0876	23	21	159	47	11	29
22. Calliope	2.9090	1801.0	0.0193	62	43	4	47	1	45
155. Scylla	2.9127	1815.7	0.2559	82	1	42	52	14	4
238. Hypatia.........	2.9163	1819.0	0.0946	32	18	184	26	12	28
231. Vindobona......	2.9192	1821.7	0.1537	253	23	352	49	5	10

Table II.—*Continued.*

Name	a	P	e	π		Ω		i	
16. Psyche	2.9210	1823.4 d	0.1392	15°	9′	150°	36′	3°	4′
179. Clytemnestra...	2.9711	1870.6	0.1133	355	39	253	13	7	47
239. Adrastea	2.9736	1873.0	0.2279	26	1	181	34	6	4
69. Hesperia	2.9779	1877.0	0.1712	108	19	187	12	8	28
150. Nuwa	2.9785	1877.5	0.1307	355	27	207	35	2	9
61. Danaë	2.9855	1884.2	0.1615	344	4	334	11	18	14
117. Lomia	2.9907	1889.1	0.0229	48	46	349	39	14	58
35. Leucothea	2.9923	1890.6	0.2237	202	25	355	49	8	12
263. Dresda	3.0120	1909.3	0.3051	308	49	217	56	1	27
221. Eos	3.0134	1910.7	0.1028	330	58	142	35	10	51
162. Laurentia	3.0241	1920.8	0.1726	145	52	38	15	6	4
156. Xantippe	3.0375	1933.7	0.2637	155	58	246	11	7	29
241. Germania	3.0381	1934.0	0.1013	340	7	272	28	5	30
256. Walpurga	3.0450	1940.8	0.1180	240	17	183	35	12	44
211. Isolda	3.0464	1942.2	0.1541	74	12	265	29	3	51
96. Ægle	3.0497	1945.3	0.1405	163	10	322	50	16	7
257. Silesia	3.0572	1952.5	0.2555	54	16	34	31	4	41
133. Cyrene	3.0578	1953.0	0.1398	247	13	321	8	7	14
95. Arethusa	3.0712	1965·9	0.1447	32	58	244°	17	12	54
202. Chryseis	3.0777	1972.1	0.0959	129	46	137	47	8	48
268.	3.0852	1973.9	0.1285	184	48	121	53	2	25
100. Hecate	3.0904	1984.3	0.1639	308	3	128	12	6	23
49. Pales	3.0908	1984.7	0.2330	31	15	290	40	3	8
223. Rosa	3.0940	1987.9	0.1186	102	48	49	0	1	59
52. Europa	3.0955	1988.0	0.1098	106	57	129	40	7	27
245. Vera	3.0985	1992.1	0.1950	25	29	62	37	5	10
86. Semele	3.1015	1995.1	0.2193	29	10	87	45	4	47
159. Æmilia	3.1089	2002.2	0.1034	101	22	135	9	6	4
48. Doris	3.1127	2005.9	0.0649	70	33	184	55	6	31
196. Philomela	3.1137	2006.8	0.0118	309	19	73	24	7	16
130. Electra	3.1145	2007.7	0.2132	20	34	146	6	22	57
212. Medea	3.1157	2008.8	0:1013	56	18	315	16	4	16
120. Lachesis	3.1211	2014.0	0.0475	214	0	342	51	7	1
181. Eucharis	3.1226	2015.4	0.2205	95	25	144	45	18	38
62. Erato	3.1241	2016.9	0.1756	39	0	125	46	2	12
222. Lucia	3.1263	2019.0	0.1453	258	2	80	11	2	11
137. Meliboea	3.1264	2019.1	0.2074	307	58	204	22	13	22
165. Loreley	3.1269	2019.6	0.0734	223	50	304	6	10	12
251. Sophia	3.1315	2024.1	0.1243	77	7	157	6	10	20
24. Themis	3.1357	2028.1	0.1242	144	8	35	49	0	49
152. Atala	3.1362	2028.6	0.0862	84	23	41	29	12	12
10. Hygeia	3.1366	2029.1	0.1156	237	2	285	38	3	49
259. Aletheia	3.1369	2029.3	0.1176	241	45	88	32	10	40
227. Philosophia	3.1393	2031.6	0.2131	226	23	330	52	9	16
147. Protogenea	3.1393	2031.6	0.0247	25	38	251	16	1	54
171. Ophelia	3.1432	2035.4	0.1168	143	59	101	10	2	34
209. Dido	3.1436	2035.9	0.0637	257	33	2	0	7	15
31. Euphrosyne	3.1468	2039.0	0.2228	93	26	31	31	26	27
90. Antiope	3.1475	2039.7	0.1645	301	15	71	29	2	17
104. Clymene	3.1507	2042.7	0.1579	59	32	43	32	2	54

Table II.—*Continued.*

Name	a	P	e	π	Ω		i	
57. Mnemosyne ...	3.1510	2043.0 d	0.1145	53° 25′	200°	2′	15°	12′
250. Bettina	3 1524	2044.3	0.1302	87 28	26	12	12	54
252. Clementina.....	3.1552	2047.1	0.0837	355 8	208	19	10	2
94. Aurora	3.1602	2052.0	0.0827	48 46	4	9	8	4
106. Dione	3.1670	2058.6	0.1788	25 57	63	14	4	38
199. Byblis............	3.1777	2069.0	0.1687	261 20	89	52	15	22
99. Undina.	3.1851	2076.3	0.1024	331 27	102	52	9	57
184. Deiopea	3.1883	2079.4	0.0725	169 22	336	18	1	12
176. Idunna.	3.1906	2081.6	0.1641	20 34	201	13	22	31
154. Bertha...........	3.1976	2088.5	0.0788	190 47	37	35	20	59
108. Hecuba..........	3 2113	2101.0	0.1005	173 49	352	17	4	24
122. Gerda............	3.2177	2108.2	0.0415	203 45	178	43	1	36
168. Sibylla	3.3765	2266.2	0.0707	11 26	209	47	4	33
225. Henrietta.......	3.4007	2277.8	0.2661	299 13	200	45	20	45
229. Adelinde	3.4129	2302.9	0.1562	332 7	30	49	2	11
76. Freia.	3.4140	2304.1	0.1700	90 49	212	5	2	3
260. Huberta.........	3.4212	2311.5	0.1113	313 22	168	48	6	18
65. Maximiliana...	3.4270	2317.2	0.1097	260 36	158	50	3	29
121. Hermione.......	3.4535	2344.2	0.1255	357 50	76	46	7	36
87. Sylvia............	3.4833	2374.5	0.0922	333 48	75	49	10	55
107. Camilla..........	3 4847	2376.0	0.0756	115 53	176	18	9	54
175. Andromache...	3.5071	2399.0	0.3476	293 0	23	35	3	46
190. Ismene	3.9471	2864.3	0.1634	105 39	177	0	6	7
153. Hilda	3.9523	2869.9	0.1721	285 47	228	20	7	55

PART II.

1. Extent of the Zone.

In Table II. the unit of column a is the earth's mean distance from the sun, or ninety-three million miles. On this scale the breadth of the zone is 1.8196. Or, if we estimate the breadth from the perihelion of Æthra (1.612) to the aphelion of Andromache (4.726), it is 3.114,—more than three times the radius of the earth's orbit. A very remarkable characteristic of the group is the interlacing or intertwining of orbits. "One fact," says D'Arrest, "seems above all to confirm the idea of an intimate relation between all the minor planets; it is, that if their orbits are figured under the form of material rings, these rings will be found so entangled that it would be possible, by means of one among them taken at hazard, to lift up all the rest."* Our present knowledge of this wide and complicated cluster is the result of a vast amount, not only of observations, but also of mathematical labor. In view, however, of the perturbations of these bodies by the larger planets, and especially by Jupiter, it is easy to see that the discussion

---

* This ingenious idea may be readily extended. The least distance of Æthra is less than the present aphelion distance of Mars; and the maximum aphelion distance of the latter exceeds the perihelion distance of several known asteroids. Moreover, if we represent the orbits of the major planets, and also those of the comets of known periods, by material rings, it is easy to see that the major as well as the minor planets are all linked together in the manner suggested by D'Arrest.

of their motions must present a field of investigation practically boundless.

While the known minor planets were but few in number the theory of Olbers in regard to their origin seemed highly probable; it has, however, been completely disproved by more recent discoveries. The breadth of the zone being now greater than the distance of Mars from the sun, it is no more probable that the asteroids were produced by the disruption of a single planet than that Mercury, Venus, the earth, and Mars originated in a similar manner.

2. The Small Mass of the Asteroids.

In taking a general view of the solar system we cannot fail to be struck by the remarkable fact that Jupiter, whose mass is much greater than that of all other planets united, should be immediately succeeded by a region so nearly destitute of matter as the zone of asteroids. Leverrier inferred from the motion of Mars's perihelion that the mass of Jupiter is at least twelve hundred times greater than that of all the planets in the asteroid ring. The fact is suggestive of Jupiter's dominating energy in the evolution of the asteroid system. We find also something analogous among the satellites of Jupiter, Saturn, and Uranus. Jupiter's third satellite, the largest of the number, is nearly four times greater than the second. Immediately within the orbit of Titan, the largest satellite of Saturn, occurs a wide hiatus, and the volume of the next interior satellite is to that of Titan in the ratio of one to twenty-one. In the Uranian system the widest interval between adjacent orbits is just within the orbit of the bright satellite, Titania.

The foregoing facts suggest the inquiry, What effect would be produced by a large planet on interior masses abandoned by a central spheroid? As the phenomena in all instances would be of the same nature, we will consider a single case,—that of Jupiter and the asteroids.

The powerful mass of the exterior body would produce great perturbations of the neighboring small planets abandoned at the solar equator. The disturbed orbits, in some cases, would thus attain considerable eccentricity, so that the matter moving in them would, in perihelion, be brought in contact with the equatorial parts of the central body, and thus become reunited with it.* The extreme rarity of the zone between Mars and Jupiter, regarded as a single ring, is thus accounted for in accordance with known dynamical laws.

3. The Limits of Perihelion Distance.

It is sufficiently obvious that whenever the perihelion distance of a planet or comet is less than the sun's radius, a collision must occur as the moving body approaches the focus of its path. The great comet of 1843 passed so near the sun as almost to graze its surface. With a perihelion distance but very slightly less, it would have been precipitated into the sun and incorporated with its mass. In former epochs, when the dimensions of the sun were much greater than at present, this falling of comets into the central orb of the system must have been a comparatively frequent occurrence. Again, if Mercury's orbit had its present eccentricity when the radius

---

* The effects of Jupiter's disturbing influence will again be resumed.

of the solar spheriod was twenty-nine million miles, the
planet at its nearest approach to the centre of its motion
must have passed through the outer strata of the central
body. In such case a lessening of the planet's mean
distance would be a necessary consequence. We thus
see that in the formation of the solar system the eccentricity of an asteroidal orbit could not increase beyond a
moderate limit without the planet's return to the solar
mass. The bearing of these views on the arrangement
of the minor planets will appear in what follows.

4. Was the Asteroid Zone originally Stable?—Distribu-
tion of the Members in Space.

One of the most interesting discoveries of the eighteenth century was Lagrange's law securing the stability
of the solar system. This celebrated theorem, however
is not to be understood in an absolute or unlimited sense
It makes no provision against the effect of a resisting
medium, or against the entrance of cosmic matter from
without. It does not secure the stability of all periodic
comets nor of the meteor streams revolving about the
sun. In the early stages of the system's development
the matter moving in unstable orbits may have been
and probably was, much more abundant than at present
But even now, are we justified in concluding that all
known asteroids have stable orbits? For the major
planets the secular variations of eccentricity have been
calculated, but for the orbits between Mars and Jupiter
these limits are unknown. With an eccentricity of
0.252 (less than that of many asteroids), the distance of
Hilda's aphelion would be greater than that of Jupiter's
perihelion. It seems possible, therefore, that certain

minor planets may have their orbits much changed by Jupiter's disturbing influence.*

Whoever looks at a table of asteroids arranged in their order of discovery will find only a perplexing mass of figures. Whether we regard their distances, their inclinations, or the forms of their orbits, the elements of the members are without any obvious connection. Nor is the confusion lessened when the orbits are drawn and presented to the eye. In fact, the crossing and recrossing of so many ellipses of various forms merely increase the entanglement. But can no order be traced in all this complexity? Are there no breaks or vacant spaces within the zone's extreme limits? Has Jupiter's influence been effective in fixing the position and arrangement of the cluster? Such are some of the questions demanding our attention. If "the universe is a book written for man's reading," patient study may resolve the problem contained in these mysterious leaves.

Simultaneously with the discovery of new members in the cluster of minor planets, near the middle of the century, occurred the resolution of the great nebula in Orion. This startling achievement by Lord Rosse's telescope was the signal for the abandonment of the nebular hypothesis by many of its former advocates. To the present writer, however, the partial resolution of a single nebula seemed hardly a sufficient reason for its summary rejection. The question then arose whether any probable test of Laplace's theory could be found in

---

* Not only nebulæ are probably unstable, but also many of the sidereal systems. The Milky Way itself was so regarded by Sir William Herschel.

the solar system itself. The train of thought was somewhat as follows: Several new members have been found in the zone of asteroids; its dimensions have been greatly extended, so that we can now assign no definite limits either to the ring itself or to the number of its planets; if the nebular hypothesis be true, the sun, after Jupiter's separation, extended successively to the various decreasing distances of the several asteroids; the eccentricities of these bodies are generally greater than those of the old planets; this difference is probably due to the disturbing force of Jupiter; the zone includes several distances at which the periods of asteroids would be commensurable with that of Jupiter; in such case the conjunctions of the minor with the major planet would occur in the same parts of its path, the disturbing effects would accumulate, and the eccentricity would become very marked; such bodies in perihelion would return to the sun, and hence blanks or chasms would be formed in particular parts of the zone. On the other hand, if the nebular hypothesis was not true, the occurrence of these gaps was not to be expected. Having thus pointed out a prospective test of the theory, it was announced with some hesitation that *those parts of the asteroid zone in which a simple relation of commensurability would obtain between the period of a minor planet and that of Jupiter are distinguished as gaps or chasms similar to the interval in Saturn's ring.*

The existence of these blanks was thus predicted in theory before it was established as a fact of observation. When the law was first publicly stated in 1866, but ten asteroids had been found with distances greater than three times that of the earth. The number of such now known is sixty-five. For more than a score of years the progress of discovery has been watched with lively interest, and the one hundred and eighty new members of the group have been found moving in harmony with this law of distribution.*

When we say that an asteroid's period is commensurable with that of Jupiter, we mean that a certain whole number of the former is equal to another whole number of the latter. For instance, if a minor planet completes two revolutions to Jupiter's one, or five to Jupiter's two, the periods are commensurable. It must be remarked, however, that Jupiter's effectiveness in disturbing the motion of a minor planet depends on the *order* of commensurability. Thus, if the ratio of the less to the greater period is expressed by the fraction $\frac{1}{2}$, where the difference between the numerator and the denominator is one, the commensurability is of the first order; $\frac{1}{3}$ is of the second; $\frac{2}{5}$, of the third, etc. The difference between the terms of the ratio indicates the frequency of conjunctions while Jupiter is completing the number of revolutions expressed by the numerator. The distance 3.277, corresponding to the ratio $\frac{1}{2}$, is the only case of the first order in the entire ring; those of the second order, answering to $\frac{1}{3}$ and $\frac{2}{5}$, are 2.50 and 3.70. These orders of commensurability may be thus arranged in a tabular form, the radius of the earth's orbit being the unit of distance :

---

* Menippe, No. 188, is placed in one of the gaps by its calculated elements; but the fact that it has not been seen since the year of its discovery, 1878, indicates a probable error in its elements.

Order.	Ratio.	Distance.
First....................	$\frac{1}{2}$	3.277
Second	$\frac{1}{3}$, $\frac{2}{5}$	2.50 3.70
Third....................	$\frac{2}{5}$, $\frac{3}{7}$, $\frac{4}{9}$	2 82 3.58 3.80
Fourth........	$\frac{3}{7}$, $\frac{5}{8}$, $\frac{7}{11}$	2.95 3 51 3.85

Do these parts of the ring present discontinuities? and, if so, can they be ascribed to a chance distribution? Let us consider them in order.

I.—The Distance 3.277.

At this distance an asteroid's conjunctions with Jupiter would all occur at the same place, and its perturbations would be there repeated at intervals equal to Jupiter's period (11.86 y.). Now, when the asteroids are arranged in the order of their mean distances (as in Table II.) this part of the zone presents a wide chasm. The space between 3.218 and 3.376 remains, hitherto a perfect blank, while the adjacent portions of equal breadth, interior and exterior, contain fifty-four minor planets. The probability that this distribution is not the result of chance is more than three hundred billions to one.

The breadth of this chasm is one-twentieth part of its distance from the sun, or one-eleventh part of the breadth of the entire zone.   •

II.—The Second Order of Commensurability.—The Distances 2.50 and 3.70.

At the former of these distances an asteroid's period would be one-third of Jupiter's, and at the latter, three-fifths. That part of the zone included between the distances 2.30 and 2.70 contains one hundred and ten intervals, exclusive of the maximum at the critical distance 2.50. This gap—between Thetis and Hestia—is not only much greater than any other of this number, but is more than sixteen times greater than their average. The distance 3.70 falls in the wide hiatus interior to the orbit of Ismene.

III.—Chasms corresponding to the Third Order.—The Distances 2.82, 3.58, and 3.80.

As the order of commensurability becomes less simple, the corresponding breaks in the zone are less distinctly marked. In the present case conjunctions with Jupiter would occur at angular intervals of 120°. The gaps, however, are still easily perceptible. Between the distances 2.765 and 2.808 we find twenty minor planets. In the next exterior space of equal breadth, containing the distance 2.82, there is but one. This is No. 188, Menippe, whose elements are still somewhat uncertain. The space between 2.851 and 2.894—that is, the part of equal extent immediately beyond the gap—contains thirteen asteroids. The distances 3.58 and 3.80 are in the chasm between Andromache and Ismene.

IV.—The Distances 2.95, 3.51,* and 3.85, corresponding to the Fourth Order of Commensurability.

The first of these distances is in the interval between Psyche and Clytemnestra; the second and third, in that exterior to Andromache.

---

* The minor planet Andromache, immediately interior to the critical distance 3.51, has elements somewhat remarkable. With

The nine cases considered are the only ones in which the conjunctions with Jupiter would occur at less than five points of an asteroid's orbit. Higher orders of commensurability may perhaps be neglected. It will be seen, however, that the distances 2.25, 2.70, 3.03, and 3.23, corresponding to the ratios of the fifth order, $\frac{2}{7}$, $\frac{3}{8}$, $\frac{4}{9}$, and $\frac{6}{11}$, still afford traces of Jupiter's influence. The first is in the interval between Augusta and Feronia; the last falls in the same gap with 3.277; and the second and third are in breaks less distinctly marked. It may also be worthy of notice that the rather wide interval between Prymno and Victoria is where ten periods of a minor planet would be equal to three of Jupiter. The distance of Medusa is somewhat uncertain.

The FACT of the existence of well-defined gaps in the designated parts of the ring has been clearly established. But the theory of probability applied in a single instance gives, as we have seen, but one chance in 300,000,000,-000 that the distribution is accidental. This improbability is increased many millions of times when we include all the gaps corresponding to simple cases of commensurability. We conclude, therefore, that those discontinuities cannot be referred to a chance arrangement. What, then, was their physical cause? and what has become of the eliminated asteroids?

---

two exceptions, Æthra (132) and Istria (183), it has the greatest eccentricity (0.3571),—nearly equal to that of the comet 1867 II. at its last return. Its perihelion distance is 2.2880, its aphelion 4.7262; hence the distance from the perihelion to the aphelion of its orbit is greater than its least distance from the sun, and it crosses the orbits of all members of the group so far as known; its least distance from the sun being considerably less than the aphelion of Medusa, and its greatest exceeding the aphelion of Hilda.

What was said in regard to the limits of perihelion distance may suggest a possible answer to these interesting questions. The doctrine of the sun's gradual contraction is now accepted by a majority of astronomers. According to this theory the solar radius at an epoch not relatively remote was twice what it is at present. At anterior stages it was 0.4, 1.0, 2.0,* etc. At the first mentioned the comets of 1843 and 1668, as well as several others, could not have been moving in their present orbits, since in perihelion they must have plunged into the sun. At the second, Encke's comet and all others with perihelia within Mercury's orbit would have shared a similar fate. At the last named all asteroids with perihelion distances less than two would have been re-incorporated with the central mass. As the least distance of Æthra is but 1.587, its orbit could not have had its present form and dimensions when the radius of the solar nebula was equal to the aphelion distance of Mars (1.665).

It is easy to see, therefore, that in those parts of the ring where Jupiter would produce extraordinary disturbance the formation of chasms would be very highly probable.

5. Relations between certain Adjacent Orbits.

The distances, periods, inclinations, and eccentricities of Hilda and Ismene, the outermost pair of the group, are very nearly identical. It is a remarkable fact, however, that the longitudes of their perihelia differ by almost exactly 180°. Did they separate at nearly the

---

* The unit being the sun's distance from the earth.

same time from opposite sides of the solar nebula?
Other adjacent pairs having a striking similarity between their orbital elements are Sirona and Ceres, Fides
and Maia, Fortuna and Eurynome, and perhaps a few
others.   Such coincidences can hardly be accidental.
Original asteroids, soon after their detachment from the
central body, may have been separated by the sun's un-
equal attraction on their parts.   Such divisions have
occurred in the world of comets, why not also in the
cluster of minor planets?

6. The Eccentricities.

The least eccentric orbit in the group is that of Philo-
mela (196); the most eccentric that of Æthra (132).
Comparing these with the orbit of the second comet of
1867 we have

The eccentricity of			Philomela $= 0.01$
"	"	"	Æthra $= 0.38$
"	"	" Comet II. 1867 (ret. in 1885)	$= 0.41$

The orbit of Æthra, it is seen, more nearly resembles
the last than the first.   It might perhaps be called
the connecting-link between planetary and cometary
orbits.

The average eccentricity of the two hundred and
sixty-eight asteroids whose orbits have been calculated
is 0.1569.   As with the orbits of the old planets, the
eccentricities vary within moderate limits, some increas-
ing, others diminishing.   The average, however, will
probably remain very nearly the same.   An inspection
of the table shows that while but one orbit is less
eccentric than the earth's, sixty-nine depart more from

the circular form than the orbit of Mercury. These eccentricities seem to indicate that the forms of the asteroidal orbits were influenced by special causes. It may be worthy of remark that the eccentricity does not appear to vary with the distance from the sun, being nearly the same for the interior members of the zone as for the exterior.

7. The Inclinations.

The inclinations in Table II. are thus distributed :

```
From  0° to  4°................................................ 70
  "   4° to  8°................................................ 83
  "   8° to 12°................................................ 59
  "  12° to 16°................................................ 32
  "  16° to 20°................................................  8
  "  20° to 24°................................................  8
  "  24° to 28°................................................  7
  "  28° to 32°................................................  0
above 32° ....................................................  1
```

One hundred and fifty-four, considerably more than half, have inclinations between 3° and 11°, and the mean of the whole number is about 8°,—slightly greater than the inclination of Mercury, or that of the plane of the sun's equator. The smallest inclination, that of Massalia, is 0° 41', and the largest, that of Pallas, is about 35°. Sixteen minor planets, or six per cent. of the whole number, have inclinations exceeding 20°. Does any relation obtain between high inclinations and great eccentricities? These elements in the cases named above are as follows:

Asteroid.	Inclination.	Eccentricity.
Pallas	34° 42′	0.238
Istria	26 30	0.353
Euphrosyne	26 29	0 228
Anna	25 24	0.263
Gallia	25 21	0.185
Æthra	25 0	0.380
Eukrate	24 57	0.236
Eva	24 25	0.347
Niobe	23 19	0.173
Eunice	23 17	0.129
Electra	22 55	0.208
Idunna	22 31	0.164
Phocea	21 35	0.255
Artemis	21 31	0.175
Bertha	20 59	0.085
Henrietta	20 47	0.260

This comparison shows the most inclined orbits to be also very eccentric; Bertha and Eunice being the only exceptions in the foregoing list. On the other hand, however, we find over fifty asteroids with eccentricities exceeding 0.20 whose inclinations are not extraordinary. The dependence of the phenomena on a common cause can, therefore, hardly be admitted. At least, the forces which produced the great eccentricity failed in a majority of cases to cause high inclinations.

8. Longitudes of the Perihelia.

The perihelia of the asteroidal orbits are very unequally distributed; one hundred and thirty-six—a majority of the whole number determined—being within the 120° from longitude 290° 50′ to 59° 50′. The maximum occurs between 30° and 60°, where thirty-five perihelia are found in 30° of longitude.

9. Distribution of the Ascending Nodes.

An inspection of the column containing the longitudes of the ascending nodes, in Table II., indicates two well-marked maxima, each extending about sixty degrees, in opposite parts of the heavens.

I. From 310° to 10°, containing 61 ascending nodes.
II.   "   120° to 180°,   "   59   "   "

Making in 120°.................... 120   "   "

A uniform distribution would give 89. An arc of 84° —from 46° to 130°—contains the ascending nodes of all the old planets. This arc, it will be noticed, is not coincident with either of the maxima found for the asteroids.

10. The Periods.

Since, according to Kepler's third law, the periods of planets depend upon their mean distances, the clustering tendency found in the latter must obtain also in the former. This marked irregularity in the order of periods is seen below.

Between 1100 and 1200 days.................. 6 periods.
  "   1200 " 1300 " .... 7 "
  "   1300 " 1400 " .... 43 "
  "   1400 " 1500 " .... 13 "
  "   1500 " 1600 " .... 46 "
  "   1600 " 1700 " .... 54 "
  "   1700 " 1800 " .... 20 "
  "   1800 " 1900 " .... 13 "
  "   1900 " 2000 " .... 19 "
  "   2000 " 2100 " .... 33 "
  "   2100 " 2200 " .... 2 "
  "   2200 " 2300 " .... 2 "
  "   2300 " 2400 " .... 8 "
  "   2400 " 2800 " .... 0 "
  "   2800 " 2900 " .... 2 "

The period of Hilda (153) is more than two and a half times that of Medusa (149). This is greater than the ratio of Saturn's period to that of Jupiter. The maximum observed between 2000 and 2100 days corresponds to the space immediately interior to chasm I. on a previous page, that between 1300 and 1400 to the space interior to the second, and that between 1500 and 1700 to the part of the zone within the fourth gap. The table presents quite numerous instances of approximate equality; in forty-three cases the periods differing less than twenty-four hours. It is impossible to say, however, whether any two of these periods are *exactly* equal. In cases of a very close approach two asteroids, notwithstanding their small mass, may exert upon each other quite sensible perturbations.

11. Origin of the Asteroids.

But four minor planets had been discovered when Laplace issued his last edition of the "Système du Monde." The author, in his celebrated seventh note in the second volume of that work, explained the origin of these bodies by assuming that the primitive ring from which they were formed, instead of collecting into a single sphere, as in the case of the major planets, broke up into four distinct masses. But the form and extent of the cluster as now known, as well as the observed facts bearing on the constitution of Saturn's ring, seem to require a modification of Laplace's theory. Throughout the greater part of the interval between Mars and Jupiter an almost continuous succession of small planetary masses—not nebulous rings—appears to have been abandoned at the solar equator. The entire cluster, distributed throughout a space whose outer radius exceeds the inner by more than two hundred millions of miles, could not have originated, as supposed by Laplace, in a single nebulous zone the different parts of which revolved with the same angular velocity. The following considerations may furnish a suggestion in regard to the mode in which these bodies were separated from the equator of the solar nebula.

(a) The perihelion distance of Jupiter is 4.950, while the aphelion distance of Hilda is 4.623. If, therefore, the sun once extended to the latter, the central attraction of its mass on an equatorial particle was but five times greater than Jupiter's perihelion influence on the same. It is easy to see, then, that this "giant planet" would produce enormous tidal elevations in the solar mass.

(b) The centrifugal force would be greatest at the crest of this tidal wave.

(c) Three periods of solar revolution were then about equal to two periods of Jupiter. The disturbing influence of the planet would therefore be increased at each conjunction with this protuberance. The ultimate separation (not of a ring but) of a planetary mass would be the probable result of these combined and accumulating forces.

12. Variability of Certain Asteroids.

Observations of some minor planets have indicated a variation of their apparent magnitudes. Frigga, discovered by Dr. Peters in 1862, was observed at the next opposition in 1864; but after this it could not be

found till 1868, when it was picked up by Professor
Tietjen. From the latter date its light seems again to
have diminished, as all efforts to re-observe it were
unsuccessful till 1879. According to Dr. Peters, the
change in brightness during the period of observation
in that year was greater than that due to its varying
distance. No explanation of such changes has yet been
offered. It has been justly remarked, however, that
"the length of the period of the fluctuation does not
allow of our connecting it with the rotation of the
planet."

13. The Average Asteroid Orbit.

At the meeting of the American Association for the
Advancement of Science in 1884, Professor Mark W.
Harrington, of Ann Arbor, Michigan, presented a pa-
per in which the elements of the asteroid system were
considered on the principle of averages. Two hundred
and thirty orbits, all that had then been determined,
were employed in the discussion. Professor Harrington
supposes two planes to intersect the ecliptic at right
angles; one passing through the equinoxes and the
other through the solstices. These planes will intersect
the asteroidal orbits, each in four points, and "the mean
intersection at each solstice and equinox may be considered a point in the average orbit."

In 1883 the Royal Academy of Denmark offered its
gold medal for a statistical examination of the orbits of
the small planets considered as parts of a ring around
the sun. The prize was awarded in 1885 to M. Sved-
strup, of Copenhagen. The results obtained by these
astronomers severally are as follows :

	Harrington.	Svedstrup.
Longitude of perihelion	14° 39′	101° 48′
" of ascending node.......	113 56	133 27
Inclination	1 0	6 6
Eccentricity...............................	0 0448	0 0281
Mean distance.......................	2.7010	2.6435

These elements, with the exception of the first, are in reasonable harmony.

14. The Relation of Short-Period Comets to the Zone of Asteroids.

Did comets originate within the solar system, or do they enter it from without? Laplace assigned them an extraneous origin, and his view is adopted by many eminent astronomers. With all due respect to the authority of great names, the present writer has not wholly abandoned the theory that some comets of short period are specially related to the minor planets. According to M. Lehmann-Filhès, the eccentricity of the third comet of 1884, before its last close approach to Jupiter, was only 0.2787.* This is exceeded by that of twelve known minor planets. Its mean distance before this great perturbation was about 4.61, and six of its periods were nearly equal to five of Jupiter's,—a commensurability of the first order. According to Hind and Krueger, the great transformation of its orbit by Jupiter's influence occurred in May, 1875. It had previously

* Annuaire, 1886.

been an asteroid too remote to be seen even in perihelion. This body was discovered by M. Wolf, at Heidelberg, September 17, 1884. Its present period is about six and one-half years.

The perihelion distance of the comet 1867 II. at its return in 1885 was 2.073; its aphelion is 4.897; so that its entire path, like those of the asteroids, is included between the orbits of Mars and Jupiter. Its eccentricity, as we have seen, is little greater than that of Æthra, and its period, inclination, and longitude of the ascending node are approximately the same with those of Sylvia, the eighty-seventh minor planet. In short, this comet may be regarded as an asteroid whose elements have been considerably modified by perturbation.

It has been stated that the gap at the distance 3.277 is the only one corresponding to the first order of commensurability. The distance 3.9683, where an asteroid's period would be two-thirds of Jupiter's, is immediately beyond the outer limit of the cluster as at present known; the mean distance of Hilda being 3.9523. The discovery of new members beyond this limit is by no means improbable. Should a minor planet at the mean distance 3.9683 attain an eccentricity of 0.3—and this is less than that of eleven now known—its aphelion would be more remote than the perihelion of Jupiter. Such an orbit might not be stable. Its form and extent might be greatly changed after the manner of Lexell's comet. Two well-known comets, Faye's and Denning's, have periods approximately equal to two-thirds of Jupiter's. In like manner the periods of D'Arrest's and Biela's comets correspond to the hiatus at 3.51, and that of 1867 II. to that at 3.277.

Of the thirteen telescopic comets whose periods correspond to mean distances within the asteroid zone, all have direct motion; all have inclinations similar to those of the minor planets; and their eccentricities are generally less than those of other known comets. Have these facts any significance in regard to their origin?

APPENDIX.

NOTE A.

THE POSSIBLE EXISTENCE OF ASTEROIDS IN UNDIS-
COVERED RINGS.

If Jupiter's influence was a factor in the separation of planetules at the sun's equator, may not similar clusters exist in other parts of our system? The hypothesis is certainly by no means improbable. For anything we know to the contrary a group may circulate between Jupiter and Saturn; such bodies, however, could not be discovered—at least not by ordinary telescopes—on account of their distance. The Zodiacal Light, it has been suggested, may be produced by a cloud of indefinitely small particles related to the planets between the sun and Mars. The rings of Saturn are merely a dense asteroidal cluster; and, finally, the phenomena of luminous meteors indicate the existence of small masses of matter moving with different velocities in interstellar space.

NOTE B.

THE ORIGIN AND STRUCTURE OF COSMICAL RINGS.

The general theory of cosmical rings and of their arrangement in sections or clusters with intervening chasms may be briefly stated in the following propositions:

I.

Whenever the separating force of a primary body on a secondary or satellite is greater than the central attraction of the latter on its superficial stratum, the satellite, if either gaseous or liquid, will be transformed into a ring.

EXAMPLES.—Saturn's ring, and the meteoric rings of April 20, August 10, November 14, and November 27. See Payne's *Sidereal Messenger*, April, 1885.

II.

When a cosmical body is surrounded by a ring of considerable breadth, and has also exterior satellites at such distances that a simple relation of commensurability would obtain between the periods of these satellites and those of certain particles of the ring, the disturbing influence of the former will produce gaps or intervals in the ring so disturbed.

See "Meteoric Astronomy," Chapter XII.; also the *Proceedings of the American Philosophical Society*, October 6, 1871 ; and the *Sidereal Messenger* for February, 1884; where the papers referred to assign a physical cause for the gaps in Saturn's ring.

THE. END.